MATHEMATICS OF THE AMERICAS

CONTENTS

The First Americans	2
On the Road	4
Cities and Villages	6
Keeping Records	8
Picture Numbers	10
All Tied Up	12
Knotty Problems	14
Art and Design	16
Surprising Symmetry	18
Mayan Numbers	20
Numbers Up and Down	22
Sports and Games	24
Counters and Dice	26
Catching the Sun	28
Sun Circles	30
On a Grand Scale	32
Pyramid of the Moon	34
Counting the Days	36
One Day, Two Names	38
Glossary and Index	40

JAMES BURNETT
CALVIN IRONS

The First Americans

The first people to live in America travelled across an ice bridge from Asia more than 15 000 years ago. Over the centuries, their descendants moved south and east and developed many different cultures. Some people of Central and South America, such as the Maya, Aztec and Inca, formed powerful states.

Early people of the Americas often made masks for special ceremonies. This mask from Peru is made from beaten gold.

These temples were burial grounds for the Maya kings of Tikal, in what is now northern Guatemala.

This timeline shows approximate dates for some events in the history of the Americas. It also shows the periods when the Maya, Aztec and Inca states were at their peak.

- 500 B.C. — Large villages built in Alaska
- A.D. — Rise of the city of Teotihuacan (Mexico)
- 500 — Foundation of Cahokia (Illinois)
- First towns built at Chaco Canyon (New Mexico)
- 1000 — Building of the Castillo at Chichen Itza (Mexico)

Maya state | Aztec state | Inca state

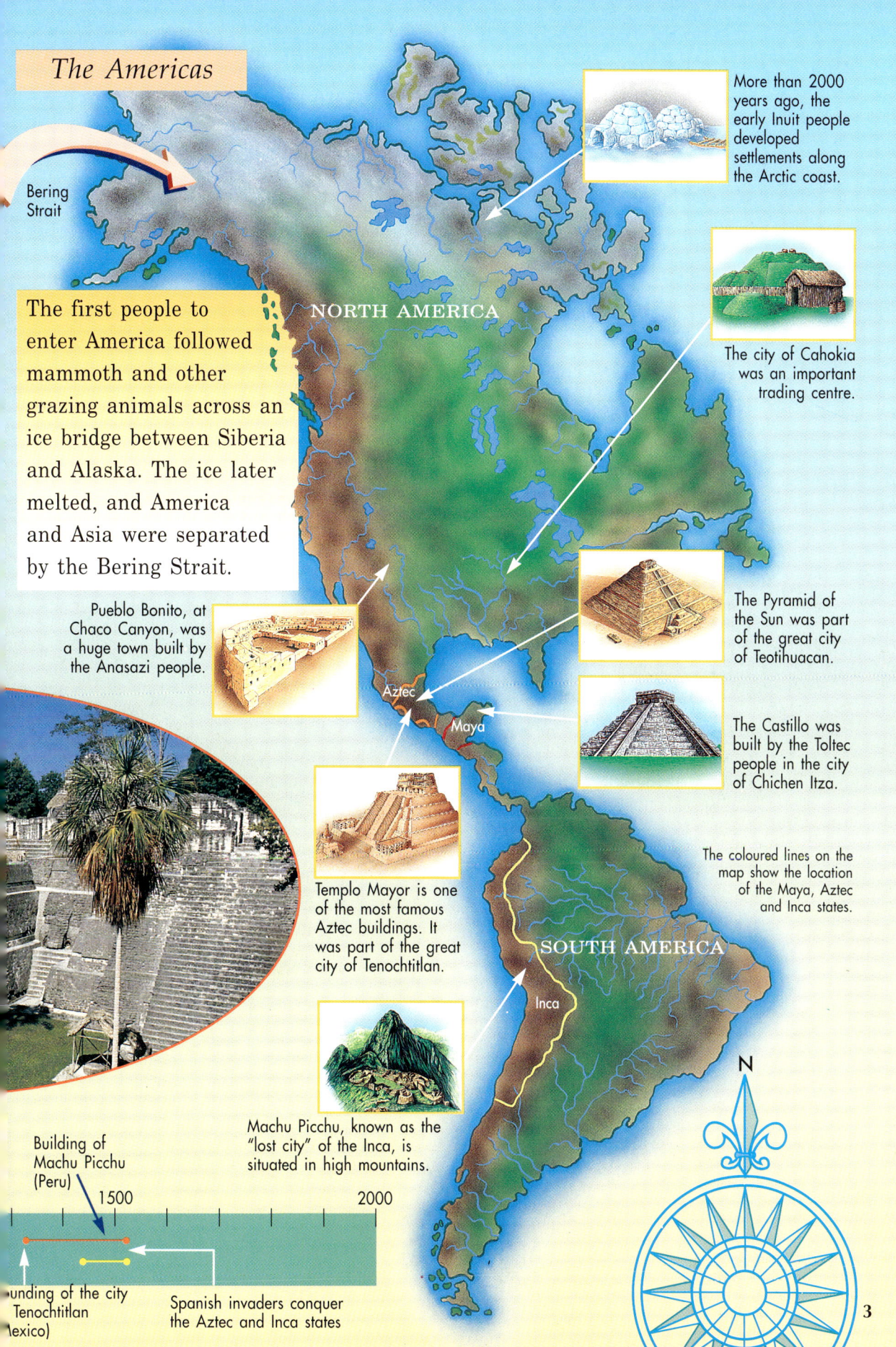

The Americas

More than 2000 years ago, the early Inuit people developed settlements along the Arctic coast.

The city of Cahokia was an important trading centre.

The first people to enter America followed mammoth and other grazing animals across an ice bridge between Siberia and Alaska. The ice later melted, and America and Asia were separated by the Bering Strait.

Pueblo Bonito, at Chaco Canyon, was a huge town built by the Anasazi people.

The Pyramid of the Sun was part of the great city of Teotihuacan.

The Castillo was built by the Toltec people in the city of Chichen Itza.

Templo Mayor is one of the most famous Aztec buildings. It was part of the great city of Tenochtitlan.

The coloured lines on the map show the location of the Maya, Aztec and Inca states.

Machu Picchu, known as the "lost city" of the Inca, is situated in high mountains.

Building of Machu Picchu (Peru)

Founding of the city of Tenochtitlan (Mexico)

Spanish invaders conquer the Aztec and Inca states

On the Road

There have been major road systems in the Americas for more than a thousand years. Some of the early roads stretched for hundreds of kilometres and were as wide as two-lane highways.

The Anasazi people of Chaco Canyon, in the present-day U.S. State of New Mexico, built over 650 km of roads. These connected more than 70 settlements. All the roads were built in straight lines, rather than being made to wind *around* cliffs and canyons. This meant that in some places they were extremely steep.

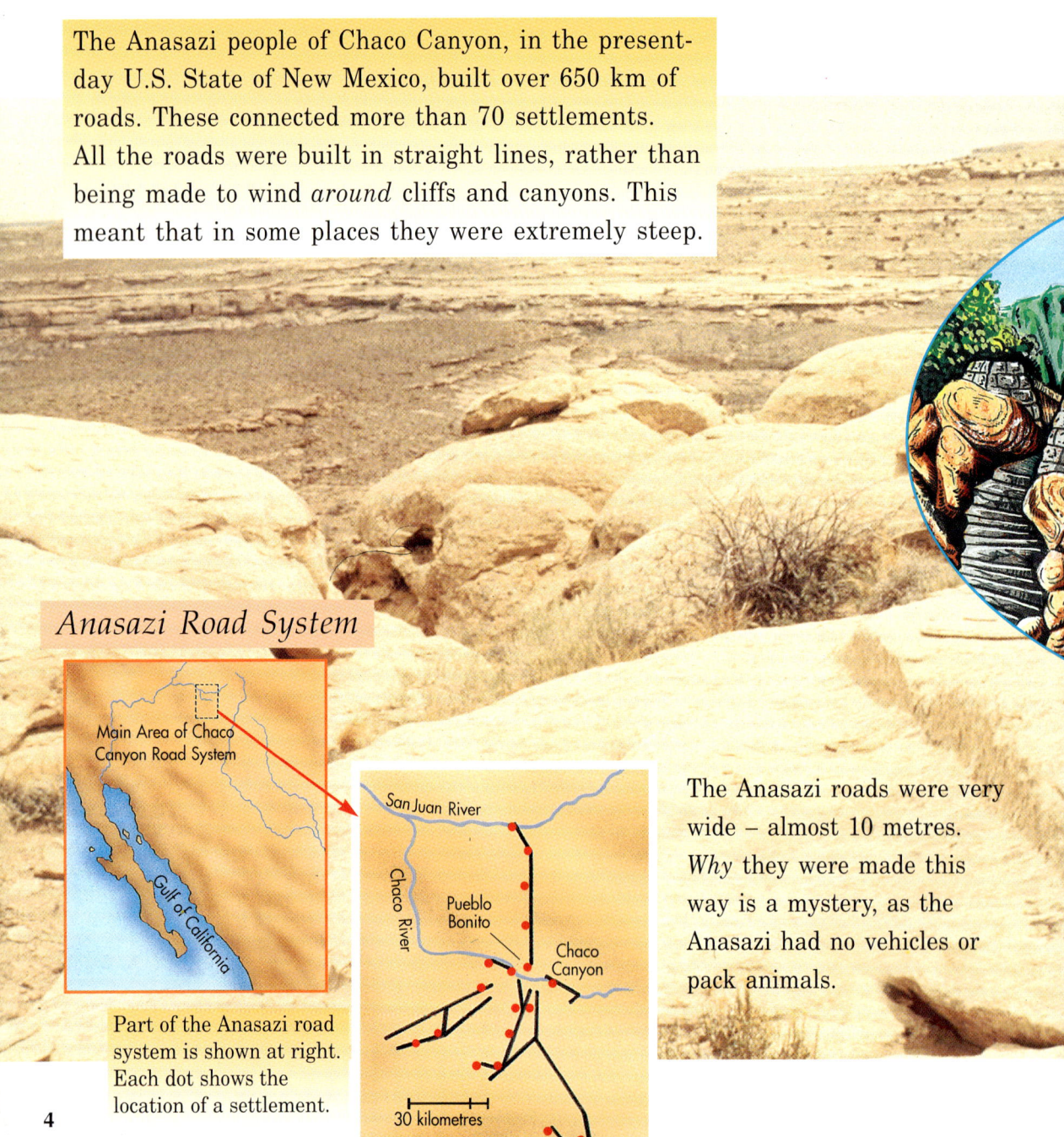

Anasazi Road System

Main Area of Chaco Canyon Road System

San Juan River

Chaco River

Pueblo Bonito

Chaco Canyon

30 kilometres

Part of the Anasazi road system is shown at right. Each dot shows the location of a settlement.

The Anasazi roads were very wide – almost 10 metres. *Why* they were made this way is a mystery, as the Anasazi had no vehicles or pack animals.

Inca Roadways

The Inca people of South America built more than 24 000 kilometres of paved roads. These varied in width from 6 metres on open ground, to 1 metre in the high mountains. The main road along the coast was more than 3900 km long. Messengers ran in relays from town to town, covering up to 400 kilometres a day.

Bridging the Gap

To cross deep gorges, the Inca built *suspension bridges* made from braided grass. The longest bridge, which crossed the 36 metre-high Apurimac Canyon, measured 66 metres in length.

Cities and Villages

The lives of the early people of the Americas varied greatly from one culture to another. Some people were *nomadic*, moving frequently to hunt for food. Others lived in villages or towns – or even great cities.

Look at pages 2–7. Answer these questions about the early people of the Americas.
1. When would people have needed to:
   a. measure things?   b. count things?
2. What other mathematics do you think they would have used?
3. Why do you think it was important for people to keep track of time?

Look at the timeline on pages 2–3.
4. Estimate when these were built:
   a. the Castillo at Chichen Itza
   b. Machu Picchu.
5. About how many years passed between the collapse of the Maya state and the rise of the Aztec state?

Time was measured in different ways by different cultures. Native Americans used notched sticks to keep track of time. They began each month with the first day of the new moon. The days were shown on the stick by a series of notches.

Research • Find out the population and area of a city near you. How does it compare to Teotihuacan?

The Aztecs did not have enough land to grow all the crops they needed, so they developed a way of using swampland for farming. Muddy soil, dug up from the swamp, was built up to create fields, or *chinampas*, separated by channels. By 1521, the chinampa system extended over 13 000 hectares.

The City of Teotihuacan

Pyramid of the Moon

The priests of Teotihuacan were skilled observers of the skies. The city was very carefully designed so that key features and buildings "lined up" with important movements of the Sun and stars.

The great Pyramid of the Sun rises more than 60 metres and measures over 210 metres along its base.

Did you know?
No one knows who built Teotihuacan. Around A.D. 750, much of the city was destroyed by fire.

More than 75 temples lined the main roadway.

Temple of the Feathered Serpent

This map shows the central area of Teotihuacan, the first planned city in the Americas. It was built around A.D. 150, and by A.D. 500 covered an area of about 20 square kilometres. Up to 200 000 people lived and worked within the city boundaries.

Keeping Records

The Aztecs developed a system of "number pictures" to help them keep accurate records. Their system used four symbols and usually involved multiples of twenty.

Did You Know?
The Aztecs counted most things using groups of 20. However, round or oval objects, such as oranges and eggs, were counted in tens.

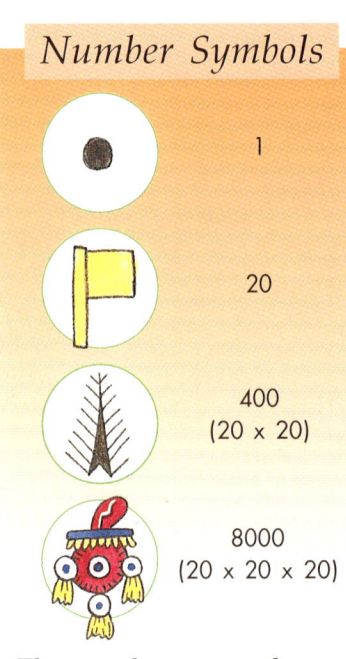

Number Symbols

- 1
- 20
- 400 (20 x 20)
- 8000 (20 x 20 x 20)

- The *number 1* was shown as a dot, representing a seed of maize.
- 20 was shown by a flag.
- 400 was shown by a feather.
- 8000 was shown by a pouch.
- There was no symbol for zero.
- There was no place-value system; number symbols could be shown in any order.

At the Market

Look for the baskets of *maize*, or corn, in this picture of an Aztec market. The Aztec symbol for 1 was a picture of a seed of maize.

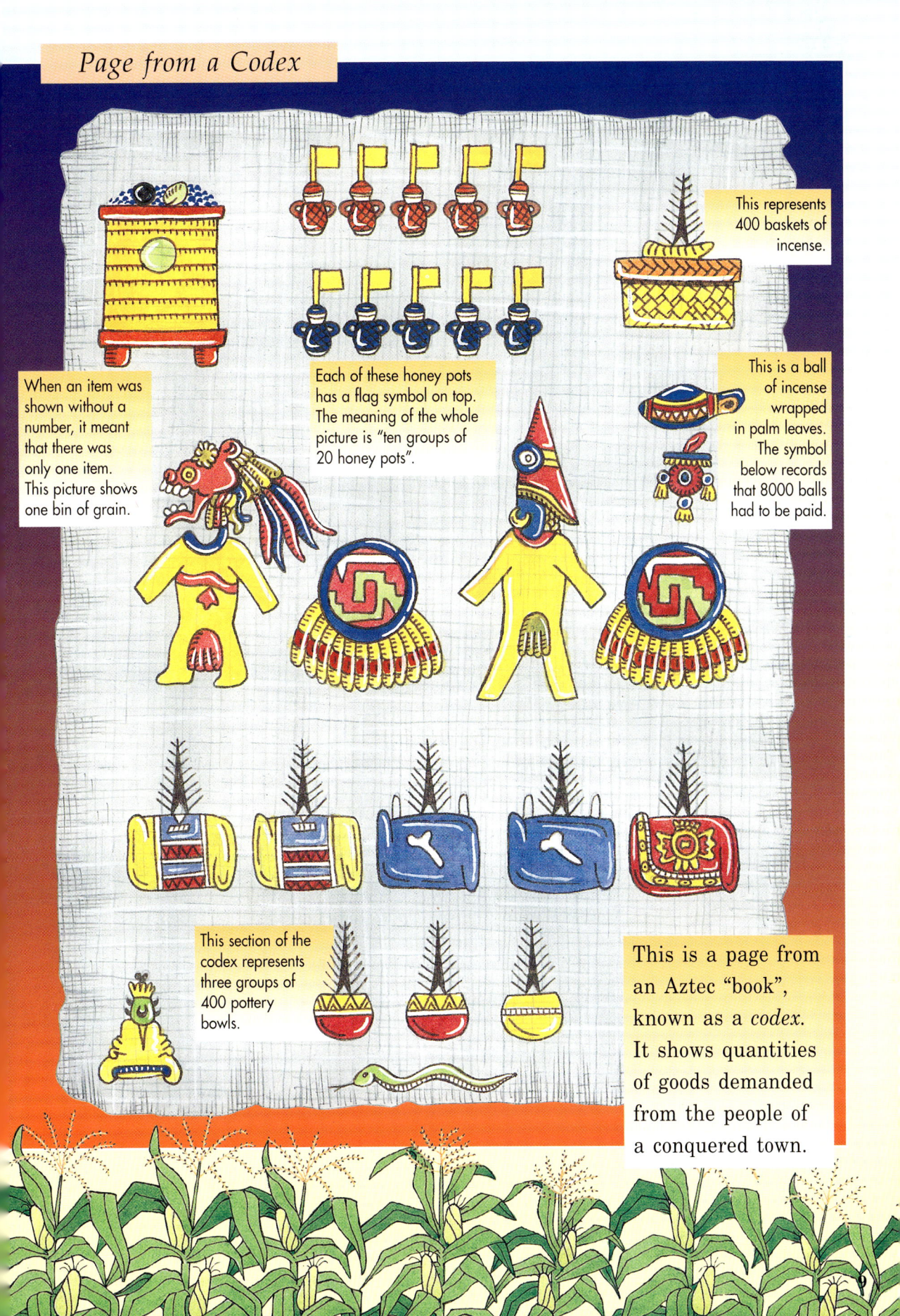

Picture Numbers

Aztec number symbols could be grouped together to show large numbers. For example, 10 of the symbols for 400 would show 4000.

Number symbols were combined with other pictures to show large quantities of particular items. This picture shows 4000 bales of cotton.

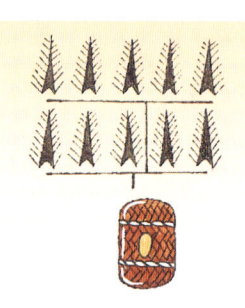

Codex Page

Key

Pottery bowl

Blanket

Axe head

Ball of incense

Bin of grain

Green feathers

Gold disk

Pot of honey

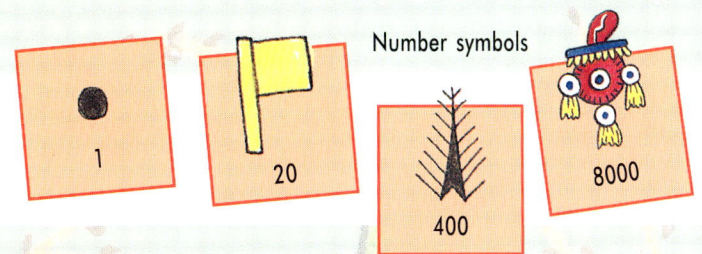

1. Look at the codex page and the key.
   a. What does each picture on the codex show?
   b. Create your own pictures to show quantities of goods. Give them to a friend to "decode".
2. How would the Aztecs have shown these numbers?
   a. 160   b. 1600   c. 16 000.
   What pattern do you see?
3. What numbers are shown below?

   a.

   b.

   c.

4. Draw Aztec symbols to show:
   a. 1040   b. 2143   c. 8602.

Research • Find out about the Egyptian number system. Compare it to the Aztec number system. List similarities and differences.

ALL TIED UP

The Inca had no system of writing numbers. Instead, they used the knotted cords of a *quipu* to show numbers. Quipus were used to record numbers describing many things; for example, population, supplies and taxes.

Lots of Knots

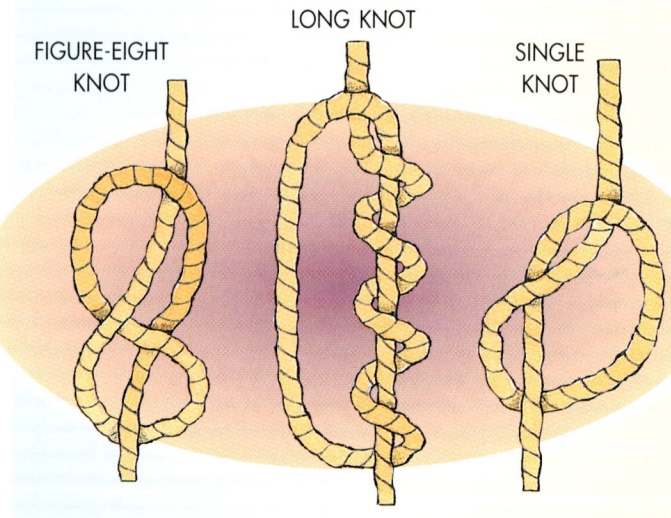

- A *figure-eight* knot is used only in the ones "place" and represents 1.
- *Long* knots with different numbers of "loops" are used in the ones place to show 2–9. The long knot above, with 4 loops, shows 4.
- *Single* knots are used for all digits in other places. For example, 3 tens are shown by three single knots in the tens place.
- There is *no knot* to show zero.

How a Quipu Works

This picture shows how a quipu is used to show numbers and to add them. Each cord of the quipu shows a different number. The last cord shows the total of the other three numbers.

This man is a modern-day quipu maker from Peru. This quipu has small objects tied into it.

hundreds place

tens place

ones place

1. This cord shows the number 150, with one single knot in the hundreds place and 5 single knots in the tens place. There is nothing in the ones place.

2. This cord shows the number 45, with 4 single knots in the tens place and a long knot with 5 loops in the ones place.

3. This cord has a figure-eight knot in the ones place. How can you tell that the whole cord shows 231?

4. This cord, looped through the other cords, shows the *sum* of the other three numbers. Check that the total is correct.

Knotty Problems

Ancient quipus were made from cotton or wool. Many involved a very complicated system of knots and could only be made and understood by experts. These experts, called *quipu camayocs*, were highly respected officials in charge of all record keeping.

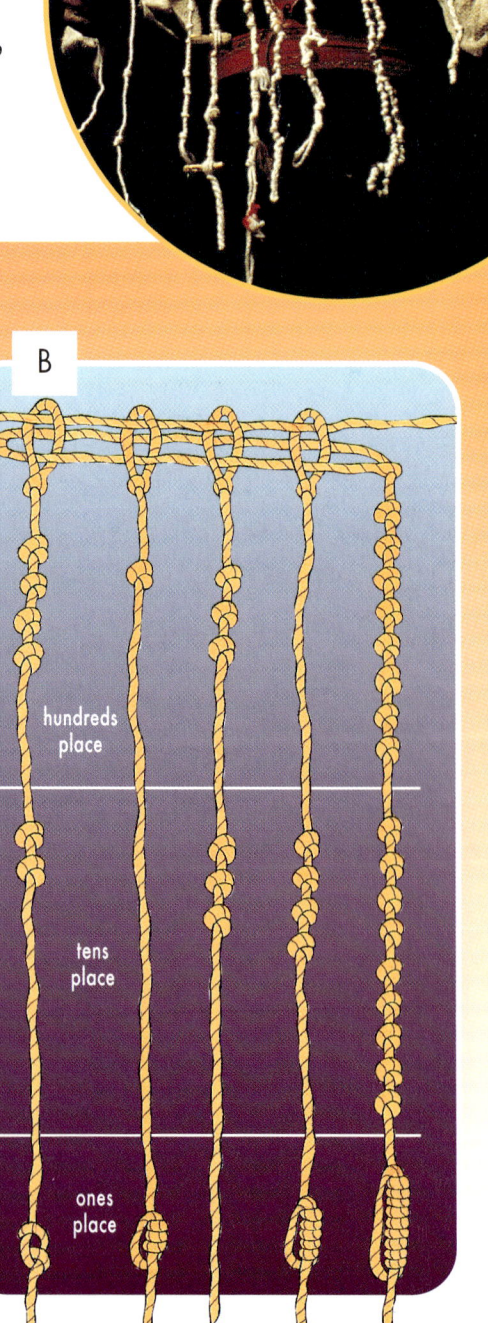

A

thousands place

hundreds place

tens place

ones place

B

hundreds place

tens place

ones place

Look at diagrams A and B. (These are simplified drawings of ancient quipus.)

1. How many numbers are shown on:
   a. quipu A?   **b.** quipu B?
2. How many places are shown on the cords of:
   a. quipu A?   **b.** quipu B?
3. Read and write the numbers shown by both quipus.
   a. Which places needed a zero?
   b. Were the numbers shown by the two "totals" cords correct?
4. Compare the method of showing numbers on a quipu with our way of writing numbers.
   How are the methods the same/different?
5. Draw your own quipu to show these numbers:

   | 241 | 403 | 165 |

   Draw another cord on your quipu to show the sum of the three numbers.
6. Make or draw a quipu to show your age, or some other number that is important to you.

Research
- Find out about other number systems that have no zero.
- Find out the correct name of the number system that we use today.

Art and Design

Early people of the Americas produced a huge variety of arts and crafts, ranging from everyday objects to valuable jewellery and pieces used in religious ceremonies. Beautifully balanced designs are a feature of these early arts and crafts.

A Precious Gift

This double-headed serpent was worn as a pendant. It is covered in precious turquoise, and was a gift from an Aztec king to a Spanish explorer about 500 years ago.

Around and Around

People of many Native American cultures wove patterned baskets. In these baskets from south-east Alaska, the designs feature rotational symmetry.

Old Gold

This ear ornament, crafted around 650 B.C., is one of the oldest gold pieces made in the Americas. It was found in Peru.

Painted Pottery

The Anasazi people of the American southwest were producing beautifully decorated pots over 1000 years ago. Many pots were decorated with geometric designs that were repeated around the top.

Surprising Symmetry

There are several different kinds of symmetry. Some designs are made to create a symmetrical *pattern*, but are not coloured symmetrically. This surprising kind of symmetry is sometimes known as *anti-symmetry*.

1. Look at the pictures on pages 16–17. What different kinds of symmetry can you see?
2. Look at the Aztec shield. Suppose you could rotate the light-coloured section to cover the darker section. Through how many degrees would it need to be rotated?
3. Follow the steps on page 19 to create a design with surprising symmetry.
4. Experiment to create a design that has the same kind of surprising symmetry as the Aztec shield. Use dot paper to help.

This Aztec warrior shield shows anti symmetry. The light and dark areas are symmetrical shapes but the colours are "opposite".

 • Look at some logos. Try to find different kinds of symmetry, including some examples of "surprising symmetry".

Create Your Own Design

You will need:
> a piece of heavy paper, no more than 8 cm x 8 cm;
> a long piece of white paper, at least 10 cm x 40 cm;
> scissors; a ruler.

Step 1: Fold the heavy paper in half. Then draw a right-angled triangle. (The fold is one side of the triangle.)

Step 2: Cut out the triangle and unfold the paper to reveal a larger triangle.

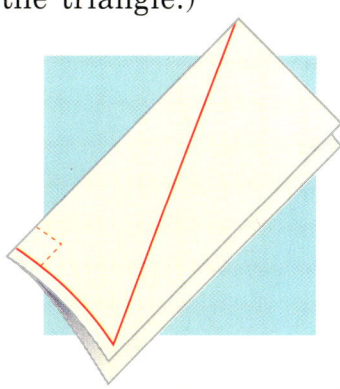
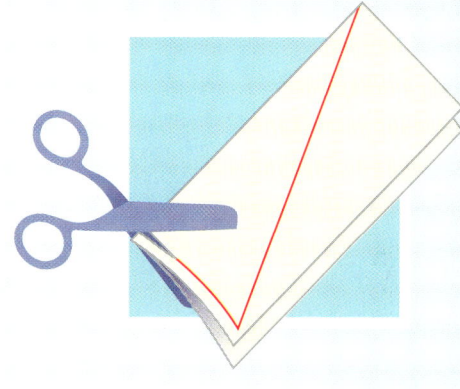

Step 3: Trace the cut-out onto the white paper as shown. Draw parallel lines to mark the borders of your design. Then draw a line at the tip of the triangle between the parallel lines.

Step 4: "Flip" the cut-out and trace it again. Continue this repeating pattern along the strip.

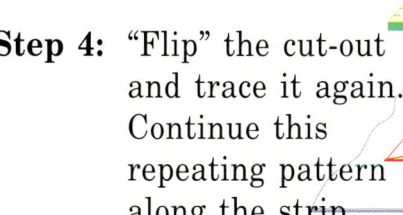

Step 5: Colour the sections alternately to create a design with *anti-symmetry*.

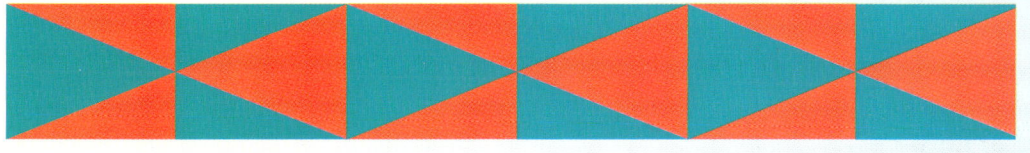

Mayan Numbers

The Mayan number system had just three symbols – a dot, a bar and a shell. The Maya were the first people to use a symbol to represent zero.

Number Symbols

The Mayan symbol for zero looked like a shell. Dots represented ones and bars represented fives.

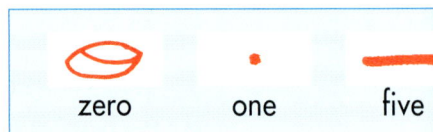

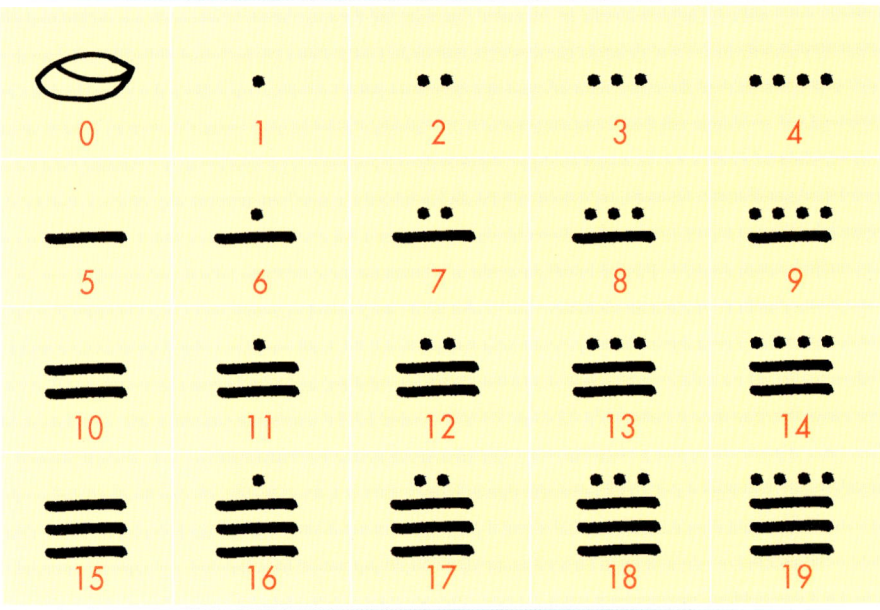

Place Value

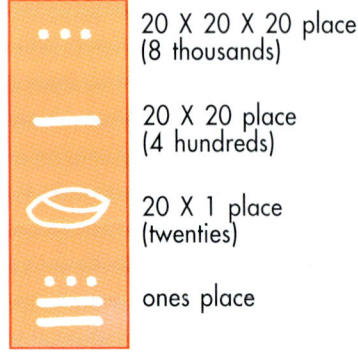

Unlike our place-value system, which is based on 10, the Mayan system is based on 20. While our numbers are arranged horizontally, Mayan numbers are *vertical*. The picture shows the number 26 013. You can find it by adding 13 ones, zero 20s, five 400s and three 8000s: (13 + 0 + 2000 + 24 000 = 26 013).

Keeping Records

Mayan priests kept records in "books" known as *codices*. These were made of long strips of bark paper that were folded like the one shown here.

A Special System

Most Mayan people used the place-value system described on page 20. However, priests had another, special place-value system that was used only for counting days or other units of time. In this system, the third place was 360 (or 18 x 20), because 360 is close to the number of days in a year. Here is an example of the way in which one number could have two values.

COUNTING OBJECTS		COUNTING DAYS
1 X 8000	•	**1** X 7200 (20 X 18 X 20)
10 X 400	=	**10** X 360 (18 X 20)
7 X 20	••	**7** X 20
13 X 1	•••	**13** X 1
12 153		**10 953**

Numbers Up and Down

There are only four Mayan codices in existence. One of these, known as the *Dresden Codex*, records information about the Sun, Moon and stars. It was used by priests to help them predict the timing of events such as eclipses. Because it deals with time, the codex uses the priest's system in which the place values are 1, 20, 360 and so on.

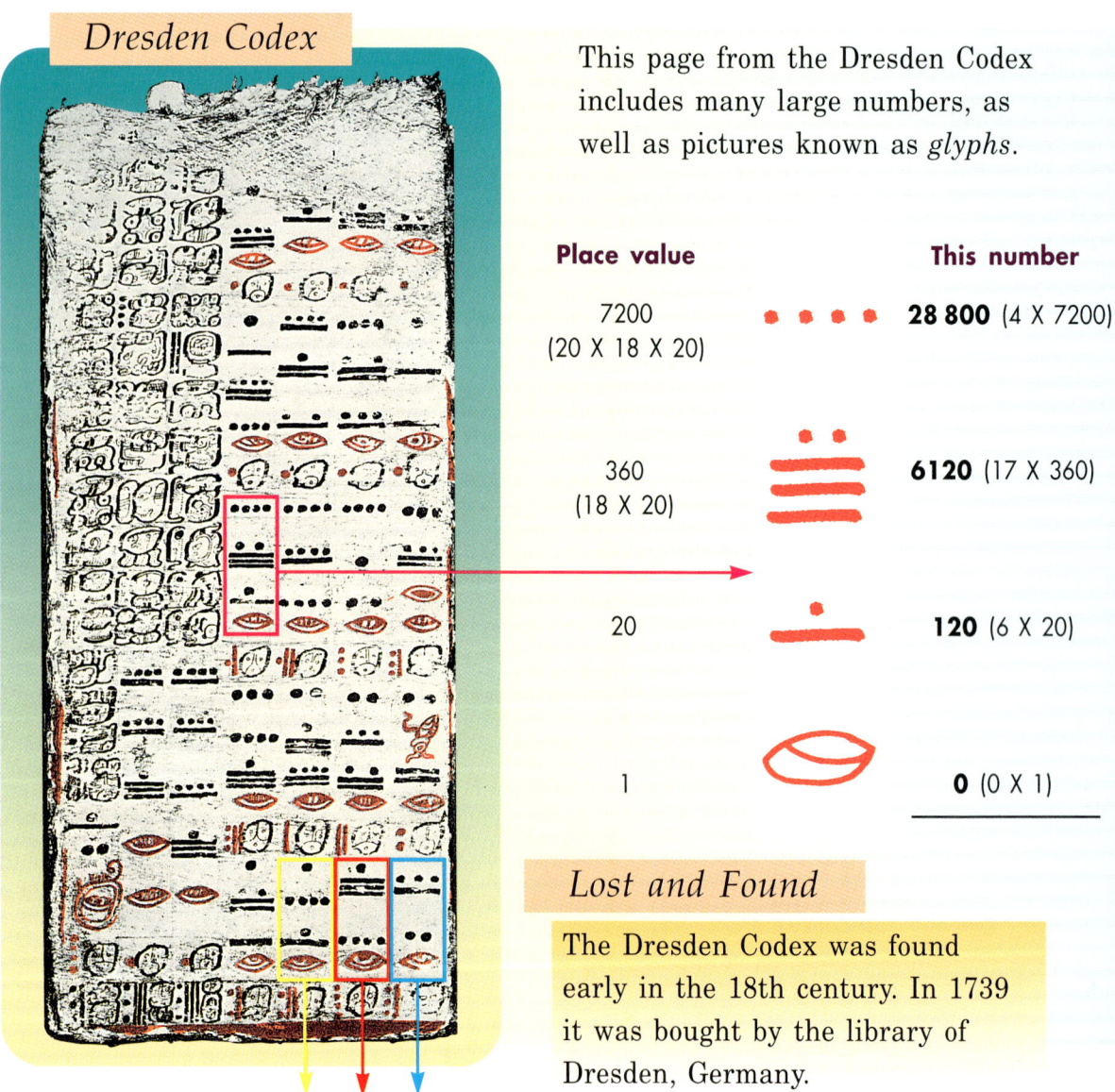

Dresden Codex

This page from the Dresden Codex includes many large numbers, as well as pictures known as *glyphs*.

Place value		This number
7200 (20 X 18 X 20)	• • • •	**28 800** (4 X 7200)
360 (18 X 20)	═ ═ ═	**6120** (17 X 360)
20	• ─	**120** (6 X 20)
1	⬭	**0** (0 X 1)

Lost and Found

The Dresden Codex was found early in the 18th century. In 1739 it was bought by the library of Dresden, Germany.

1. Look at the number in the codex that is outlined in pink.
   a. How many places does it have?
   b. What is its value?
2. Look at the numbers outlined at the bottom of the codex.
   a. Number A has 4 places. Copy the number and give its value.
   b. Numbers B and C each have 3 places. Give the values of the numbers.
   c. Describe any patterns you can see in the values of these three numbers.
3. How would a Mayan priest have written these numbers of days:
   a. 20?
   b. 25?
   c. 400?
4. Using the priest's system, what is the *greatest* number you can write using:
   a. 2 places?
   b. 3 places?
   c. 4 places?
5. How do you think it helped the Maya to have a symbol for zero?

Sports and Games

People in the Americas developed a variety of sports and games. Some of these were played as tests of stamina and strength rather than games of fun.

Fast and Furious

Tlachti was a very fast ball game, something like basketball. It was first played by ancestors of the Maya.

The largest *tlachti* ball court was at Chichen Itza. It was 149 metres long and its stone walls were 8.2 metres high. Two teams of two or three players tried to "shoot" a solid rubber ball through a stone ring set high up on the walls. This was a difficult task, as players had to hit the ball with their elbows and hips!

Patolli

Patolli was a popular Aztec board game. Players tossed "dice" made from cacao beans before moving their pebble counters.

Istaboli was one of several games played in south-eastern North America, by tribes such as the Choctow. The playing field measured about 70 metres in length, and goalposts at each end were as high as 6 metres. Teams had as many as 100 players. Although similar to modern-day lacrosse, *istaboli* was sometimes more like a savage battle than a sport.

COUNTERS AND DICE

Many people of the Americas played board games and games of chance using "dice".

Bone Dice: a Cree Game

Players toss eight "dice" that are white on one side and dark on the other. Four are hook-shaped and four are diamonds. Different combinations give different scores. The first person to score 100 points wins.

1. How could you win Bone Dice in one toss? Would this be likely to happen?
2. List some combinations that would give a score of zero.
3. Suppose you had three tosses. Write some scores that you think are:
   a. possible
   b. impossible
   c. likely
   d. unlikely.
4. List some different tosses that would give a total of 32.

Scoring Chart

100	all white sides up
80	all dark sides up
30	7 white sides and 1 dark side up
10	white sides up on all hooks plus 1 white diamond
8	dark sides up on all hooks plus 1 dark diamond
4	white sides up on all diamonds plus 1 white hook
2	all hooks standing on edge
0	any other combinatio

Tuknanavuhpi: a Hopi Game for Two Players

How to Play

The two players need twenty counters each – a different colour for each player. The aim is to jump your opponent's pieces and remove them from the board. Players cannot jump their own pieces.

1. Players arrange their counters as shown at right, leaving the centre position empty.
2. Players take turns moving a counter to a vacant position on the board. Moves can be made in any direction.
3. When a line across an end of the board becomes empty, it cannot be used again.
4. The game is finished when one player has removed all the other player's counters from the board.

Catching the Sun

In many early American cultures, special ceremonies took place on days that marked changes of season. Knowing exactly when these days would occur required careful observations of the Sun.

Dagger of Light

The Anasazi people created a device to indicate the summer solstice. A 30-centimetre wide spiral was carved into a cliff face on this steep mountain, or *butte*, in present-day New Mexico. Just before noon on the summer solstice, a spot of light appears above the spiral. During the next 20 minutes, the light stretches into a dagger shape, piercing the centre of the spiral, then slides off the bottom and disappears.

The Special Days: Equinox and Solstice

Solstices are the longest and shortest days of the year. These days occur when the Sun is the greatest distance from the equator (about June 21 and December 21). *Equinoxes* are the two days when night and day each last 12 hours. These occur when the Sun is closest to the equator (about March 21 and September 23).

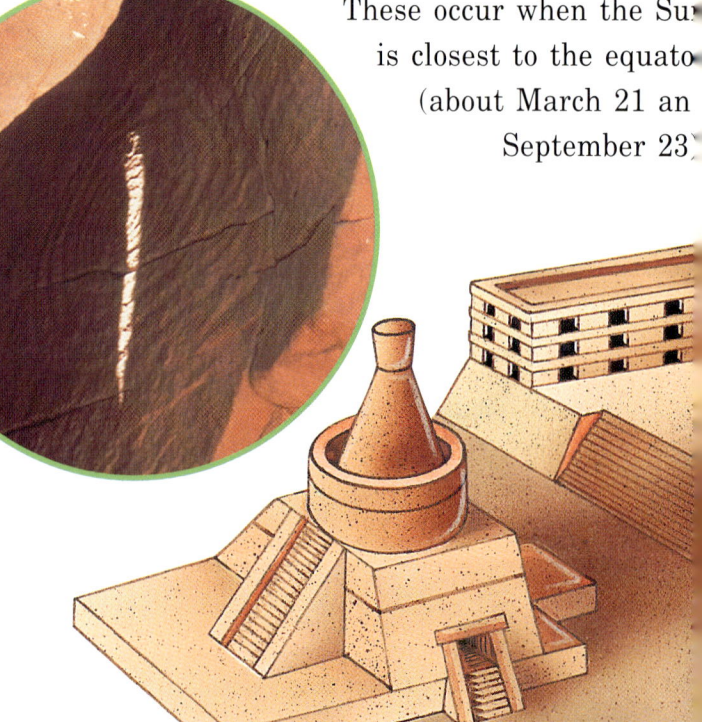

Cahokia

The large city of Cahokia, in the present-day State of Illinois, included over 120 massive earth mounds. Many of these were built to "line up" with movements of the Sun. A special "observatory" or *sun circle* was used for watching the Sun and identifying each solstice and equinox.

Aztec Sun Watchers

The Templo Mayor, a great Aztec building in present-day Mexico, was positioned so that the rising sun of each equinox would shine directly between the two temples on top of the pyramid. This event signalled the start of an important ritual.

Sun Circles

At Cahokia, special "sun circles" were built to help priests observe the Sun's position. One circle, marked out with 48 tall posts, measured 124 metres in diameter.

A sun priest would stand at the centre of the sun circle. The priest faced the rising Sun to observe its position on the horizon.

This diagram shows how 48 posts were evenly spaced, approximately 8 metres apart, to form a sun circle. The posts no longer exist, but a number of post holes were discovered in 1961.

Sun Circle

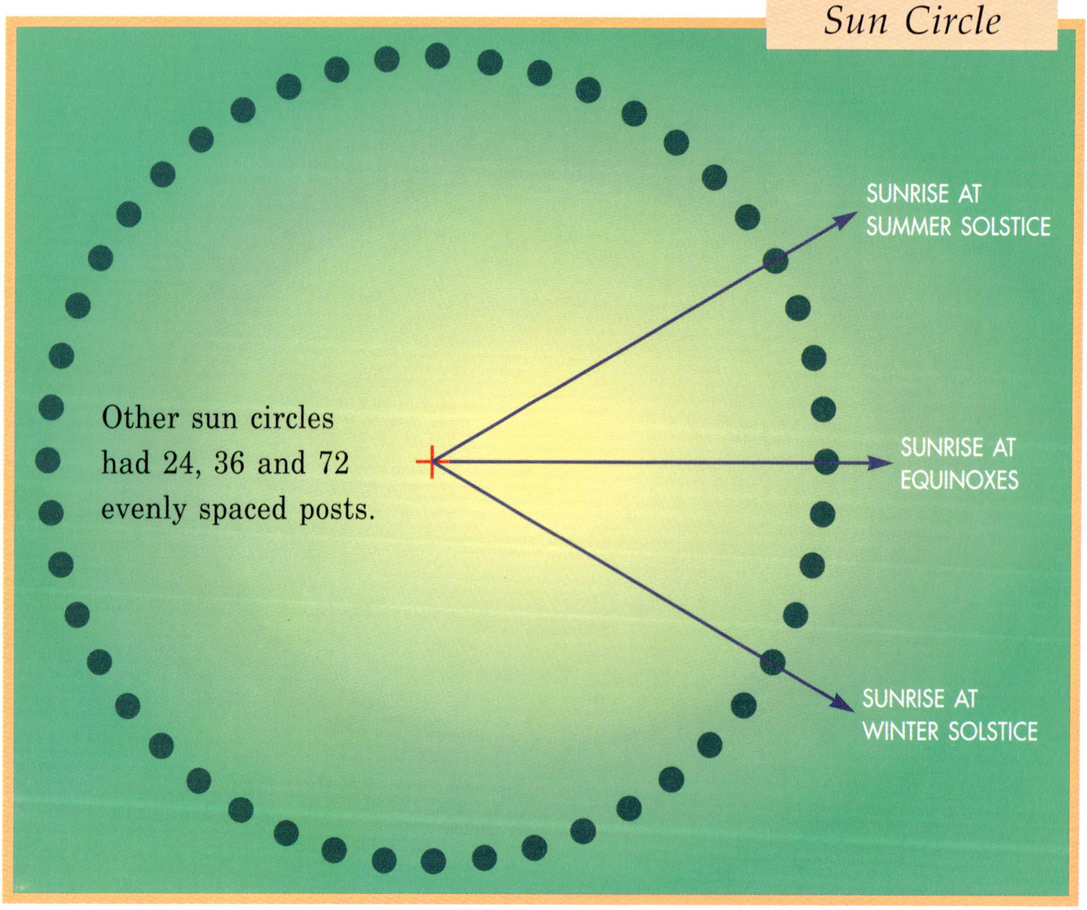

Other sun circles had 24, 36 and 72 evenly spaced posts.

SUNRISE AT SUMMER SOLSTICE

SUNRISE AT EQUINOXES

SUNRISE AT WINTER SOLSTICE

Look at the diagram of the sun circle.
1. Use a protractor.
   a. Find the size of the angle between the equinox "line" and the summer solstice line. Check that the angle between the equinox line and the winter solstice line is the same.
   b. Predict the size of the angle between the winter solstice line and the summer solstice line. How did you make your prediction?
2. Explain how you could find the angle between any post and the post next to it:
   a. by using the angles you found in question 1
   b. by using what you know about the total number of degrees in a circle.
3. Copy and complete the chart below. Explain how you found the missing numbers. Describe any patterns you see.

NUMBER OF POSTS (evenly spaced)	RADIUS	DIAMETER	CIRCUMFERENCE	ANGLE BETWEEN POSTS	DISTANCE BETWEEN POSTS
24	35 metres				
48	62 metres				
72	68 metres				
36	62 metres				

ON A GRAND SCALE

People of the Americas built many large structures – from "apartment buildings" made up of many rooms to huge pyramids. Some of these structures are still standing, after more than a thousand years.

Pueblo Bonito

These ruins at Chaco Canyon, New Mexico, are all that remain of a multi-storey Anasazi village built between A.D. 900 and A.D. 1150. It had about 800 rooms, housing more than 1200 people.

The Pyramid of Niches

This pyramid was built around A.D. 600 in Tajin, Mexico. It has a square base measuring 35 metres along each side, and once stood about 24.5 metres high. Each of its seven levels has a number of openings or *niches*. Altogether, there are 365 niches, one for each day of the year.

Castillo at Chichen Itza

The Castillo was built by the Toltec people in the ancient city of Chichen Itza. The pyramid has a square base measuring 55.5 metres along each side, and it rises to a height of 24 metres. Each face of the pyramid has a stairway of 91 steps. Together with the top level, this makes a total of 365 steps – the number of days in a year.

33

Pyramid of the Moon

The great Aztec city of Teotihuacan contained many large structures: from government buildings to palaces. One of the most impressive buildings was the Pyramid of the Moon. This was a temple used for special ceremonies.

1. Calculate the length of a *hunab* in metres and centimetres.
2. Look at the top view of the Pyramid of the Moon.
   a. Find the number 30. What does this number tell you?
   b. How wide is the base of the pyramid? Give your answer in hunabs.
   c. What is the distance along the ground from the front of the pyramid to the back of the pyramid?
3. Look at the front view.
   a. Find the number 4. What does this number tell you?
   b. Estimate the pyramid's height in metres and centimetres.
4. Use the scale and a ruler to estimate the dimensions of the "roof" of the pyramid. Give your estimate in metres and centimetres.

 • The Pyramid of the Moon covers an area of about 2.6 hectares. Compare this to the area of your school grounds.

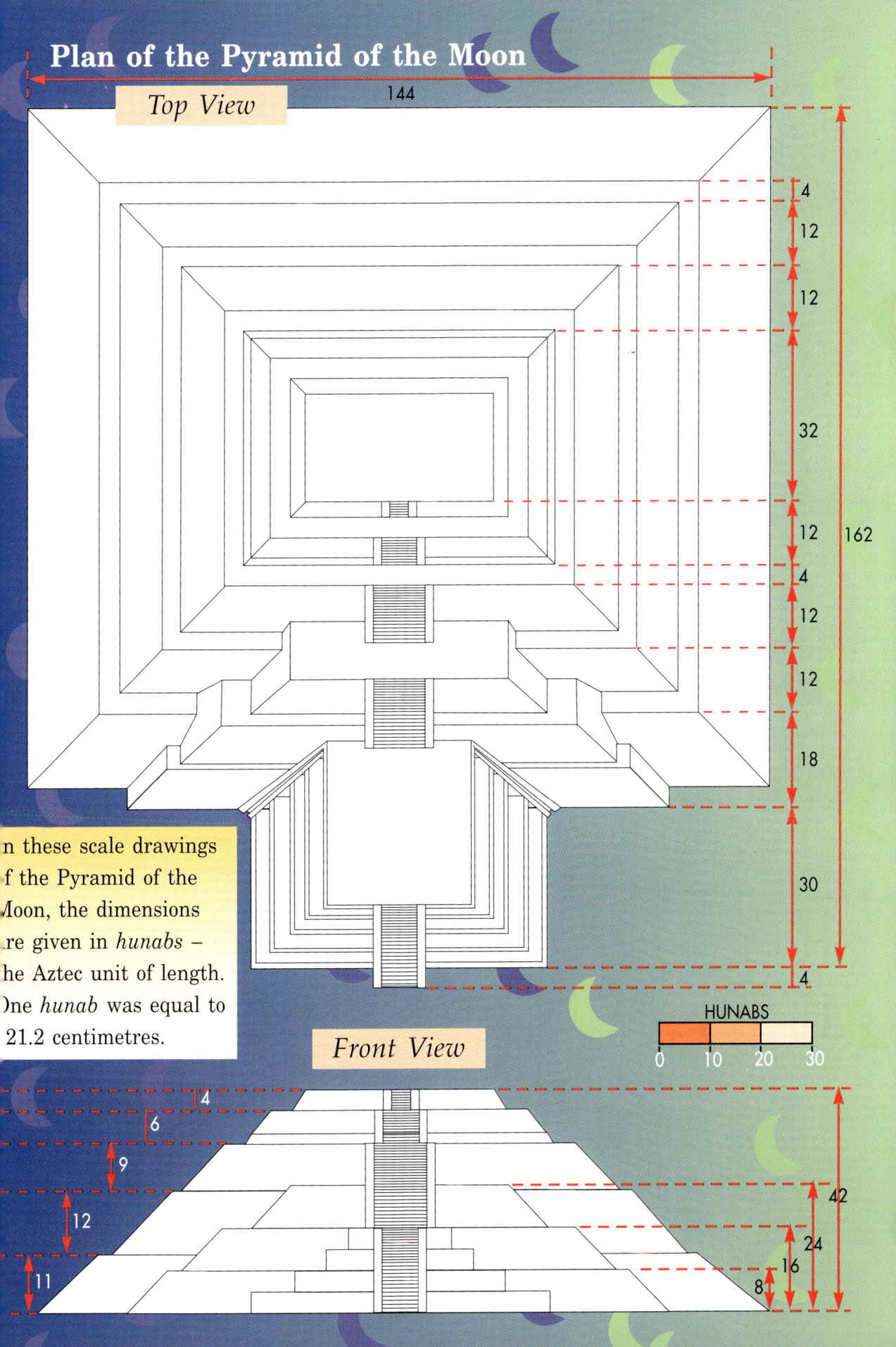

COUNTING THE DAYS

The Maya developed a complex calendar system. It was based on two separate cycles: a "sacred cycle" with 260 days, and a "solar cycle" with 365 days. The two cycles worked together, so that each day had a name from the sacred cycle *and* a name from the solar cycle.

The Sacred Cycle

The 260-day cycle used 20 days, and the numbers 1 through 13. It was called the "sacred calendar" because it was used to decide the timing of sacred ceremonies. No one knows why a cycle of 260 days was chosen.

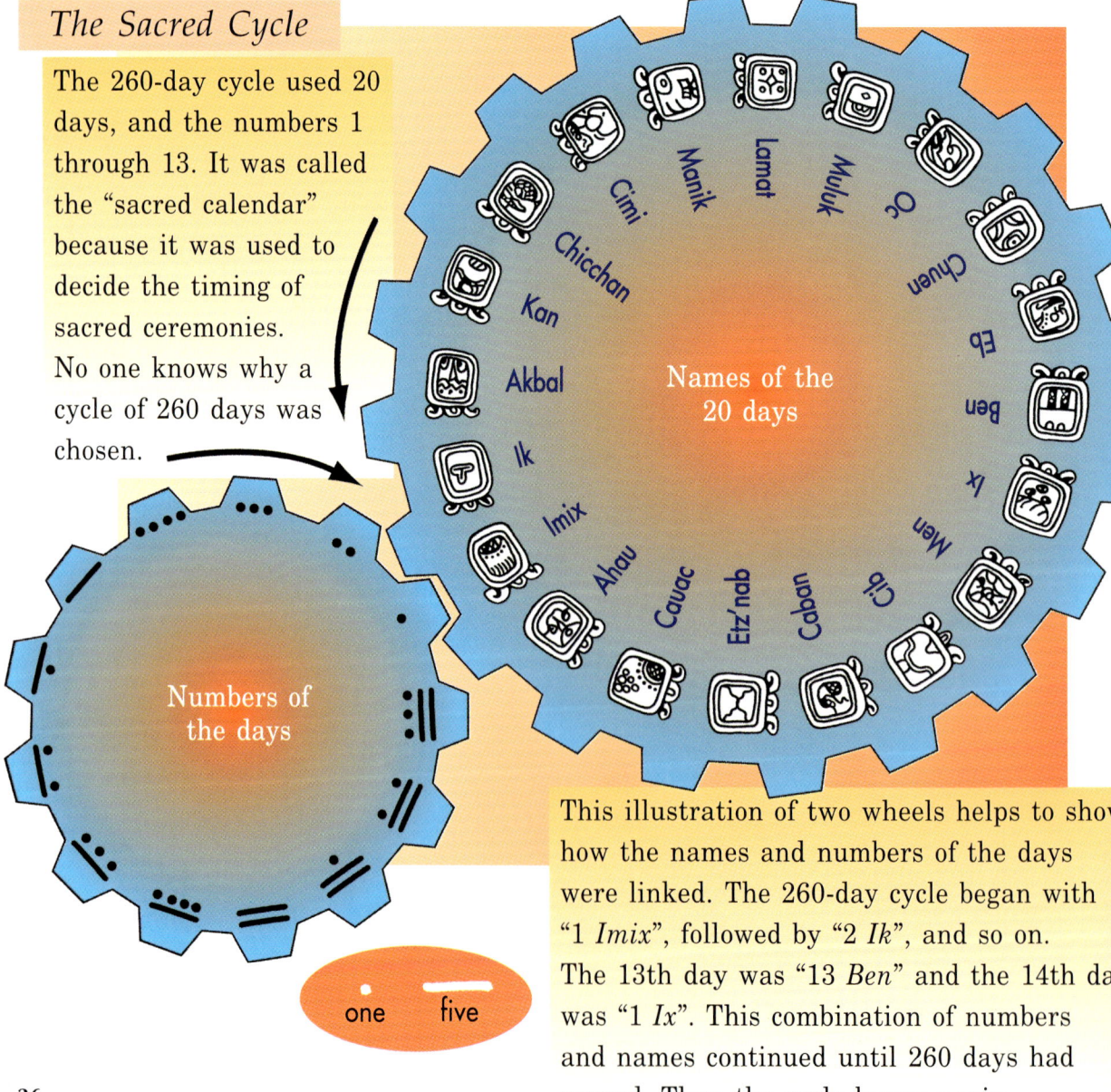

This illustration of two wheels helps to show how the names and numbers of the days were linked. The 260-day cycle began with "1 *Imix*", followed by "2 *Ik*", and so on. The 13th day was "13 *Ben*" and the 14th day was "1 *Ix*". This combination of numbers and names continued until 260 days had passed. Then the cycle began again.

Months in the Solar Cycle

The Solar Cycle

The 365-day cycle was more like our own calendar, as it was based on the natural cycles of the Earth and Sun. There were 18 months of 20 days each. In addition, there was a dreaded time of 5 unlucky days, known as *Uayeb*.

This is the lid of the stone "coffin" or *sarcophagus* of a Mayan king. Carvings on the coffin show details of the king's life. The section at right shows that he was born on "8 *Ahau* 13 *Pop*" and died on "6 *Etz'nab* 11 *Yax*".

One Day, Two Names

Each date in the Mayan calendar had a "double" name; it included a name and number from the 260-day cycle *and* a name and number from the 365-day cycle. This meant that a full name, such as "2 Ik 3 Zac", was only repeated once every 18 980 days.

Linking the Two Cycles

The "wheels" diagram opposite is one way of showing how the two cycles worked together to create a date. It is now showing "1 Imix 2 Zac". The next day will be "2 Ik 3 Zac", and so on.

Look at the "wheels" diagram.
1. Each month in the solar cycle has 20 days. How many days in the month of Zac:
   a. are shown on the diagram?
   b. are not shown on the diagram?
2. The date shown where the wheels connect is "1 Imix 2 Zac". The next day will be "2 Ik 3 Zac".
   a. What day will come after that?
   b. What day came *before* "1 Imix 2 Zac"?
3. How many of our years are equal to 18 980 days?
4. How does the Mayan solar cycle compare to our year? Make a list of similarities and differences.

Research • Find out when the *Gregorian* calendar system that we use was created. Who introduced it?

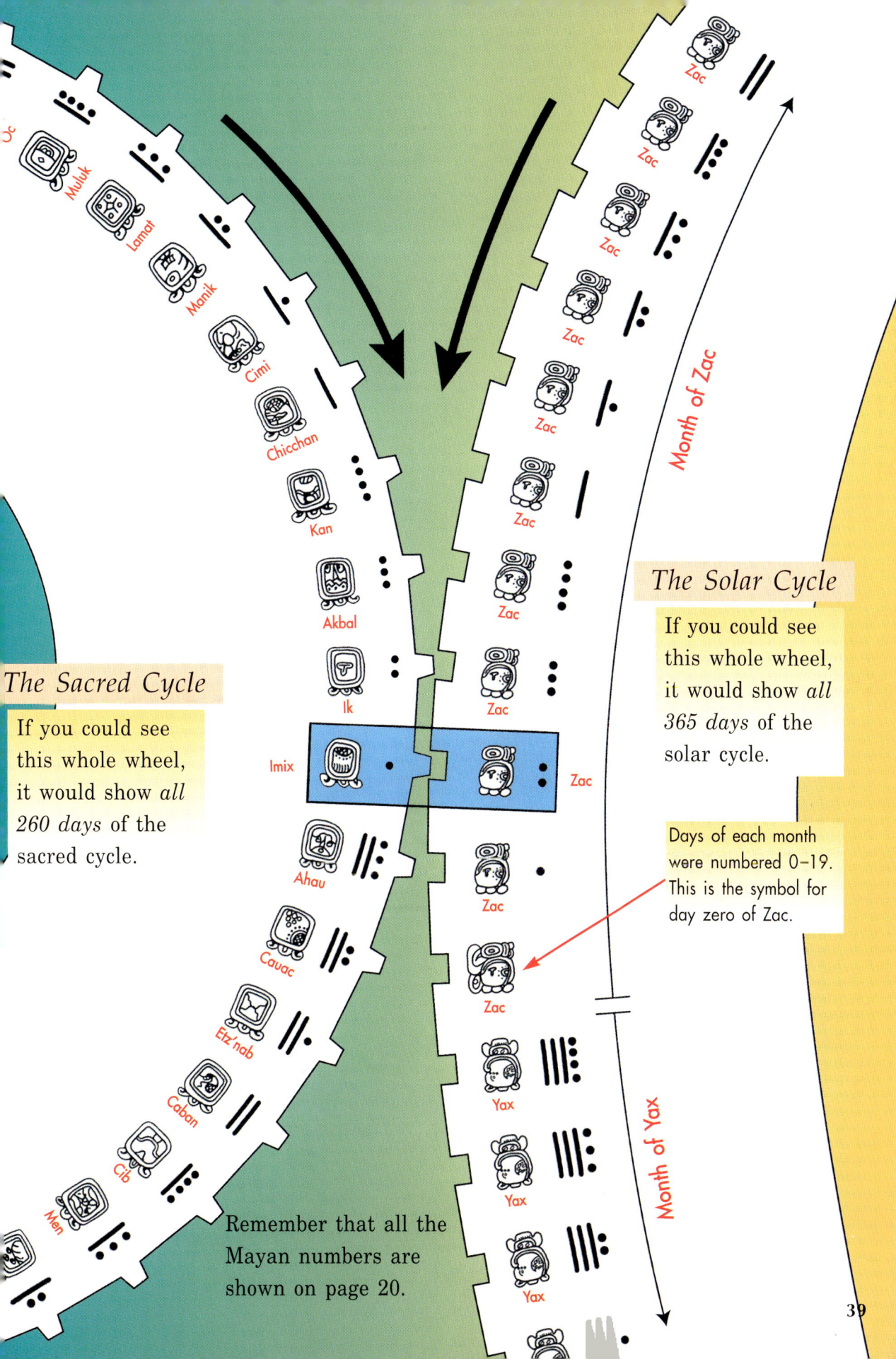

Glossary

Angle
Two straight lines form an angle when they join to make a corner. A right angle is an angle that makes a square corner (90 degrees).

Area
The amount of surface enclosed within a boundary. Area is usually measured in *square* units, such as square metres.

Circumference
The line that forms a circle, or the length of that line.

Degree
A unit used to measure the size of an angle or an amount of rotation. One degree is $\frac{1}{360}$ of one complete turn or circle.

Diameter
A line across a circle or sphere that passes through the centre and connects two points on the circumference. The distance represented by this line is also called the diameter.

Digit
The single symbols used to represent a number are known as *digits*. Our base-ten number system has ten single digits: 0, 1, 2, 3, 4, 5, 6, 7, 8, 9.

Equinox/Equinoxes
The two days of the year when night and day last 12 hours each.

Number System
An organised method of writing numbers.

Parallel Lines
Two lines or curves that are always the same distance apart.

Radius
A line from any point on the edge of a circle or sphere to its centre. The distance represented by this line is also called the radius.

Solstice/Solstices
The day of the year that has the greatest amount of sunlight is the *summer solstice*. The day that has the least amount of sunlight is the *winter solstice*.

Symmetry
a. When one side of a shape is a mirror image of the other side, the shape is said to be *symmetrical* or to have *line symmetry*.
b. When a shape can be turned through a fraction of a full rotation and still look the same, it is said to have *turning* or *rotational symmetry*.
c. When a design is symmetrical in shape but not in colour, it is said to be *anti-symmetrical*.

Index

addition 8–15, 20–23, 26, 33–35
division 32, 36–38,
geometry
  angle 30–31
  circle
    circumference 30–31, 36
    diameter 30–31
    radius 30–31
  degrees 30–31
  parallel lines 19
  patterns 16, 19
  polygons 32–33
  symmetry
    anti-symmetry 18–19
    line 16–19
    turning 16–19
  three-dimensional shapes 7, 32–34
logical thinking 27
measurement
  area 7, 34–35
  length 4–5, 7, 19, 24–25, 28, 30–35
multiplication 8–15, 20–23, 33, 36–38
number patterns 11, 20, 23, 31
numbers
  Aztec 8–11
  Egyptian 11
  Inca 12–15
  Maya 20–23
place value 8, 12–15, 20–23
probability 26
subtraction 2–3, 5–15, 24–25, 32–35
time
  days 21, 33, 36–38
  equinox 28–31
  minutes 28
  solstice 28–31
  timeline 2–3
  years 2–3, 6–7, 32–33, 36–38
visual thinking 18–19